Organic Chemistry Laboratory Notebook

This laboratory notebook belongs to

Name _____

Address _____

Telephone number _____

Course _____ Section _____

Laboratory instructor's name _____

Laboratory partner's name _____

Equipment drawer or locker number _____

Date: from_____ to _____

Copyright © 1998 Brooks/Cole, a part of Cengage Learning

Telephone Numbers of Emergency Services

Campus police _____

Campus fire _____

Health center _____

Poison control center _____

Suggestions for Using This Laboratory Notebook

1. Fill in your name, address, and telephone number on the first page so your notebook can be returned if lost.

2. Create a table of contents for your notebook as you use it during the course. Record the titles of the experiments next to the page numbers on which your write-up can be found.

3. Notebook entries are a **permanent record** of your laboratory work. Never remove the original (white) pages. The yellow (copy) pages are perforated for easy removal, if you are required to turn in a copy of your work.

4. Always **write in ink** using a firm-tipped pen, such as a ballpoint. Use enough pressure to ensure that a clear, dark copy is made.

5. If you make a mistake, draw a single line through the error and write the correct entry nearby.

6. Provide the information asked for at the top of each notebook page.

7. Use the left column to record data and calculations. In the right column write a report of your work. Include all of the elements required by your laboratory instructor, such as the purpose of the experiment, the procedure used, observations made, and conclusions drawn.

Table of Contents

51 _____

52 _____

53 _____

54 _____

55 _____

56 _____

57 _____

58 _____

59 _____

60 _____

61 _____

62 _____

63 _____

64 _____

65 _____

66 _____

67 _____

68 _____

69 _____

70 _____

71 _____

72 _____

73 _____

74 _____

75 _____

76 _____

77 _____

78 _____

79 _____

80 _____

81 _____

82 _____

83 _____

84 _____

85 _____

86 _____

87 _____

88 _____

89 _____

90 _____

91 _____

92 _____

93 _____

94 _____

95 _____

96 _____

97 _____

98 _____

99 _____

100 _____

Safety Contract*

Whenever I am in an area where laboratory reagents are being used, I agree to abide by the following rules:

1. Wear safety goggles.

2. Wear proper clothing.

3. Use good housekeeping practices.

4. Do only authorized experiments, and work only when the laboratory instructor or another qualified person is present.

5. Treat laboratory reagents as if they are poisonous and corrosive.

6. Dispense reagents carefully. Dispose of laboratory reagents as directed.

7. Not eat, drink, use tobacco, or apply cosmetics in the laboratory.

8. Report all incidents to the laboratory instructor.

9. Be familiar with the location and use of all safety equipment.

10. Become familiar with each laboratory assignment before coming to the laboratory.

11. Anticipate the common hazards that may be encountered in laboratory.

12. Become familiar with actions to be taken in the event of incidents in the laboratory.

student signature _____ date _____

laboratory instructor _____ date _____

In the space below, give any health information, such as pregnancy or other circumstance, that might help the laboratory instructor provide a safer environment for you, or that could aid the laboratory instructor in responding to an incident involving you in the laboratory.

1. I do/do not (circle one) expect to wear contact lenses during laboratory work. [Note: Goggles must still be worn when contact lenses are worn.]

2. List any known allergies to medication or other chemicals.

*From Rapp, M.W. *Practicing Safety in the Organic Chemistry Laboratory*; Chemical Education Resources: Palmyra, PA, 1997

Exp. 2: TLC & HPLC Analysis   8/6/2016

Name	Course	Section	Lab partner
Hannah Ward	211		

Names of TLC: Developing Solvent Systems:

1) 15:85 Hexane/Ethyl Acetate
2) 100% Ethyl Acetate
3) 15:85 Methanol/Ethyl Acetate

Sketch of TLC plates:

TLC R_f Calculations

1)

A $\dfrac{2cm}{5cm} = .40$

B $\dfrac{.6cm}{5cm} = 0.20$

U $\dfrac{2.7cm}{5cm} = 0.54$

C $\dfrac{2cm}{5cm} = 0.04$

D $\dfrac{2.9cm}{5cm} = 0.58$

2)

A $\dfrac{3.2cm}{5cm} = 0.64$

B $\dfrac{1.7cm}{5cm} = 0.34$

U $\dfrac{3.1cm}{5cm} = 0.62$

C $\dfrac{.6cm}{5cm} = 0.12$

D $\dfrac{3cm}{5cm} = 0.60$

3)

A $\dfrac{2.5cm}{5cm} = 0.50$

B $\dfrac{2.9cm}{5cm} = 0.58$

U $\dfrac{3.4cm}{5cm} = 0.68$

C $\dfrac{1.8cm}{5cm} = 0.36$

D $\dfrac{3.4cm}{5cm} = 0.68$

* Unknown Solution: B

Experiment title and number	Date
3: Extraction of Analgesics	9/27/16

Name	Course	Section	Lab partner
Hannah Ward	211		

Extraction:

Theoretical
Recovery
Calculation:
(ASP)

Theoretical
Recovery
Calculation:
(ACE)

Drying & Acetaminophen Recovery:

Mass 150mL beaker: 75.8 g
Mass 150mL beaker & acetaminophen: 76.14g
Recovered mass of acetaminophen: 100.4

% Recovery
Calculation: $\dfrac{76.14g}{75.8g} \times 100 = 100.4$

Aspirin Recovery:

Mass of sm. filter: 0.17g
Mass of Lg. filter: 0.71g
Mass sm. filter + Lg. filter + Aspirin: _____
Recovered mass of aspirin: _____

% Recovery
Calculation:

HPLC Analysis:

Asp vial slot #: _____
Ace vial slot #: _____

Caution: Place fold-in flap under yellow sheet before writing, to protect the pages that follow.

3: Extraction of Analgesics — 9/27/16

| Name | Course | Section | Lab partner |

Hannah Ward — 211

Extraction:

Theoretical
Recovery
Calculation:
  (ASP)

Theoretical
Recovery
Calculation:
  (ACE)

Drying & Acetaminophen Recovery:

Mass 150mL beaker: 75.8g
Mass 150mL beaker & acetaminophen: 76.14g
Recovered mass of acetaminophen: 100.4
  % Recovery
  Calculation: $\dfrac{76.14g}{75.8g} \times 100 = 100.4$

Aspirin Recovery:

Mass of sm. filter: 0.17g
Mass of Lg. filter: 0.71g
Mass sm. filter + Lg. filter + Aspirin: _____
Recovered mass of aspirin: _____
  % Recovery
  Calculation:

HPLC Analysis:

Asp vial slot #: _____
Ace vial slot #: _____

Experiment title and number	Date

Exp 4 Mutarotation of Glucose

Name	Course	Section	Lab partner
Miller Bartholomew	CHM 211	212	Andi Dai

Angle of rotation of 1 g/mL Solution: <u>49.7068°</u>

Keq for mutarotation: <u>2.03</u>

Concentration of unknown solution: <u>0.27 g/mL</u>

$$y = 48.793(1) + 0.9138 = 49.7068$$

$$49.7068 = 112.2\, C_\alpha + 18.7(1-C_\alpha)$$

$$\boxed{C_\alpha = 0.33}$$

$$\boxed{C_B = 0.67}$$

$$\frac{C_B}{C_\alpha} = 2.03$$

Concentration (g/mL)	α rotation
0.0	0°
0.1	8 → 6 } +2 blank
0.2	13 → 11 }
0.04	22°
0.6	29°
Unknown	14°

$$y = 48.793x + 0.9138$$

$$R^2 = 0.9913$$

TLC & HPLC Analysis of Analgesis 5 9/14/17

Name	Course	Section	Lab partner
Miller Bartholomew	CHM 211	212	Andi Dai + Courtney

Names of TLC Developing Solvent Systems Unknown:

① 15:85 hexane/ethyl acetate
② 100% ethyl acetate
③ 15:85 methanol/ethyl acetate

A

Sketch of TLC Plate

① ② ③

TLC Rf calculations

①

A $0.3/4.5 = .070$

B $0/4.5 = 0.0$

C $0/4.5 = 0.0$

D $1/4.5 = 0.\overline{22}$

U_1 $0.1/4.5 = .0\overline{22}$

U_2 $0.4/4.5 = .0\overline{88}$

②

A $2.8/4.7 = .59$

B_1 $1.3/4.7 = .27$
B_2 $2.2/4.7 = .46$

C $.9/4.7 = .19$

D $3.3/4.7 = .70$

U_1 $1/4.7 = .21$

U_2 $2.7/4.7 = .57$

③

A $3/4.5 = .6\overline{6}$

B $2.5/4.5 = .55$

C $1.5/4.5 = .3\overline{3}$

D_1 $2.5/4.5 = .55$
D_2 $3/4.5 = .6\overline{6}$

U_1 $1.5/4.5 = .3\overline{3}$
U_2 $3/4.5 = .6\overline{6}$

Exp. 6 Extraction of Analgesics   9/28/17

| Name | Course | Section | Lab partner |

Miller Bartholomew   CHM211 212 Andi Dai

extraction:
Theoretical recovery for aspirin: __10 g__
(1.5 X 2
(.~~828~~.

Theoretical recovery for acetaminophen: 0.65 g
     .325 X 2

drying + acetaminophen recovery
Mass of 150 mL beaker: 68.1 g
Mass of 150 mL beaker + acetaminophen: 68.53 g
actual recovered mass of acetaminophen: 0.43 g
Percent recovery for acetaminophen: ~~63~~ 66.1%

Aspirin recovery:
mass of small filter paper: 0.2 g
Mass of large filter paper: 0.67 g
mass of small filter + large filter + dried aspirin: 1.67 g
actual recovered mass of dry aspirin: 0.8 g
+Percent recovery for aspirin: __8%__

HPLC Analysis:
HPLC vial slot # for aspirin sample: __2__
HPLC vial slot # for acetaminophen sample: __1__

Experiment title and number			Date

Lab 7: Recrystallization, Melting Point, + HPLC Analysis 10/12/17

Name	Course	Section	Lab partner
Miller Bartholomew	CHM 211	212	Andi Dai

Weight of large filter paper + small filter paper + aspirin (from extraction)
$$1.67g$$

Actual weight of aspirin (from extraction): 0.80g

Weight of small filter paper: 0.18g

Weight of large filter paper (for storage): 0.71g

Weight of large filter paper + small filter paper + aspirin (from recrystalization)
$$1.54g$$

Actual weight of aspirin (after recrystalization): 0.65g

Experimental melting point range of EXTRACTED aspirin ($T_i - T_f$)
$$124 - 128°C$$

Experimental melting point range of IMPURE aspirin ($T_i - T_f$)
$$115 - 120°C$$

Experimental melting point range of RECRYSTALLIZED aspirin ($T_i - T_f$)
$$126 - 131°C$$

% recovery calculation (from recrystalization): 65%

full calc: $\dfrac{0.65g}{1.0g} = 0.65 \Rightarrow 65\%$

HPLC vial slot # for aspirin: 5

Acetaminophen

initial weight of acetaminophen (from extraction): 0.43g

weight of small filter paper: 0.16g

weight of large filter paper: 0.69g (for storage)

Weight of large filter paper + small filter paper + acetaminophen (from recrystallization)
$$1.04g$$

Actual weight of acetaminophen (after recrystallization): 0.19g

% recovery calculation (from recrystallization - full calc): 29.23%

$$\dfrac{0.19g}{0.65g} \times 100 = 29.23\%$$

Experimental melting point range of recrystallized acetaminophen 161 - 163°C

HPLC vial slot # for acetaminophen: 6

Caution: Place fold-in flap under yellow sheet before writing, to protect the pages that follow.

Exp. 8 Distillation & Gas Chromatography of alkanes 10/19/17

| Name | Course | Section | Lab partner |

Miller Bartholomew CHM211 212 And, Dai

(a) record which unknown: **B**

record Which was used

(T$_i$ - T$_f$) .80g   1.67g

record distillation range of each fraction collected: **chart**

record GC vial slot fraction #2: **5**

record GC vial slot fraction #5: **6**

(round bottom flask)

record GC sample solvent identity: **n-pentane** 1.54g

0.65g

show example adjusted area % calculation: **Below**

(full calculation)

adjusted area % (peak at 3.086 min)

$$\frac{5.439}{(5.439 + 2.634)} \times 100\% = 67.0\%$$

65%

adjusted area % (peak at 4.355 min)

$$\frac{2.634}{5.439 + 2.634} \times 100\% = 33.0\%$$

Fraction #	range °C	
	T$_i$	T$_f$
1	67°C	70°C
2	70°C	74°C
3	74°C	79°C
4	79°C	85°C
5	85°C	XXX

0.65g

1.04g

0.19g
29.23%

$$\frac{0.19g}{0.65g} = \times 100 = 29.23\%$$

08

Experiment title and number

Exp. 9 Conversion of Alcohol to Alkyl Bromide

Date 10/26/17

Name Miller Bartholomew

Course CHM 211

Section 212

Lab partner Andi Dai

initial mass of 2-methylcyclohexanol used: 5.0 g

initial volume of 48% HBr used: 10 mL

initial 50 mL beaker mass: 30.94 g

final 50 mL beaker + product mass: 36.73 g

final product mass: 5.79 g

physical state and color of product: yellow/gold liquid

theoretical yield calculation: 7.75 g product formed based on

5.0 g reactant alcohol	1 mol rct	1 mol product	177.08 g	= 7.75 g
	114.19 g	1 mol rct	1 mol product	

15.65 g alcohol product formed based on HBr

Percent yield calculation: 74.71%

Alcohol: $\dfrac{5.79 g}{7.75 g}$ = X 100 = 74.71%

Silver nitrate test reaction rates (min/s): ⟶

Test Tube | Times

1	never formed solid
2	1.1 min ⟹ 66 sec
3	2 sec
4	1 sec

10 mL 48% HBr	1.49 g	1 mol HBr	1 mol Product	177.08 g	= 32.61 g × 0.48 =
	1 mL	80.91 g	1 mol HBr	1 mol product	15.65 g

Experiment title and number		Date	
Exp.10 Acid Catalyzed Dehydration of an Alcohol w/ rearrangement		11/2/17	
Name	Course	Section	Lab partner
Miller Bartholomew	CHM 211	212	Andi Dai

initial mass of 2-methylcyclohexanol used: <u>5.03 g</u>

initial mass of 10mL round bottom flask: <u>21.75g</u>

final mass of 10mL round bottom flask + product: <u>23.42g</u>

final product mass: <u>1.67g</u>

physical state and color of product: <u>cloudy liquid</u>

GC vial slot #: <u>7</u>

GC sample solvent identity: <u>Methanol</u>

% yield calculation: <u>39.42%</u>

$$\frac{1.67g}{4.236 g} = 0.3942 \times 100\% = 39.42\%$$

Theoretical yield calculation: 4.236g

5.03 g	1 mol reactant	1 mol prod.	96.17 g
	114.19 g	1 mol reactant	1 mol product

$$\frac{actual}{theoretical} = 4.236 g$$

Show example adjusted area % calculation using the example chromatogram on the next slide

ret. time (min)	Area %
2.555	91.88721
3.086	5.43891
4.355	2.67388

adjusted area % (peak at 3.086 min) = [5.439 / (5.439 + 2.634)] ×100% = 67%

adjusted area % (peak at 4.355) = [2.634 / (5.439 + 2.634)] ×100% = 33%

min.

Exp.11 Bromination of Stilbene - Green Synthesis   11/9/17

had to leave @ 3:30 for a meeting.

Name   Course   Section   Lab partner

Miller Bartholomew   CHM 211   212   Andi Day

Initial mass of trans-stilbene used: 0.5g

Initial volumes of 48% HBr and 30% H_2O_2 used: 1 mL HBr / 1 mL H_2O_2

Mass of small filter paper: 0.19g

Mass of empty watch glass: 60.86g

Mass of watch glass + filter + product: 61.65g

Mass of final product: 0.6g

Physical state and color of product: cloudy white liquid

Theoretical yield calculations based on all 3 reactants

30% H_2O_2

1.0 mL	1.11 g	1 mol H_2O_2	1 mol product	340 g	= 11.1 g prod × 0.30 =
	1 mL	34 g	1 mol H_2O_2	1 mol prod	3.33 g

48% HBr

1.5 mL	1.49 g	1 mol HBr	1 mol prod	340 g	= 4.75 g prod × 0.48 =
	1 mL	80 g	2 mol HBr	1 mol prod	2.28 g

trans-stilbene 0.5 g	1 mol Stil.	1 mol prod	340 g	= 0.943 g
	180 g	1 mol Stil.	1 mol prod	product

Percent yield calculation: 63.83%

$$\frac{0.60 g}{0.94 g}$$

TLC diagram, including cm measurements of all spots

A: 0.89
B: 0.89
C: 0.54

A-sample
B-dibromo
C-+-stilbene

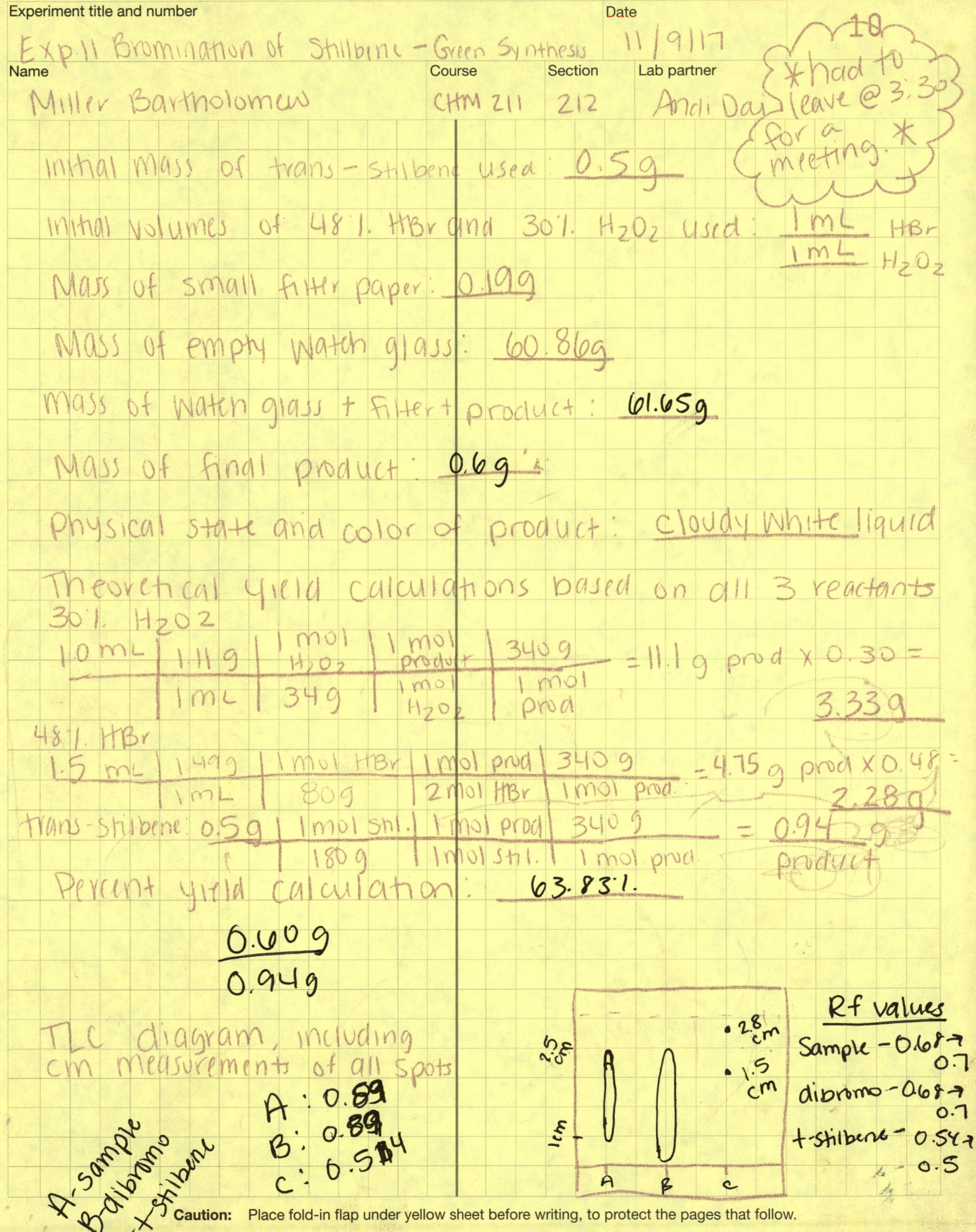

2.5 cm
•2.8 cm
•1.5 cm
1 cm
A   B   C

Rf values
Sample - 0.68 → 0.7
dibromo - 0.68 → 0.7
t-stilbene - 0.54 → 0.5

Caution: Place fold-in flap under yellow sheet before writing, to protect the pages that follow.

Experiment title and number		Date	
Lab 12: Synthesis of Diphenylacetylene		11/16/17	

Name	Course	Section	Lab partner
Miller Bartholomew	CHM 211	212	Andi Dai

Mass of dibromostilbene used: 0.47 g

Mass of KOH used: 0.42 g

Volume of ethylene glycol used: 4 mL

Volume of ethanol used: 3 mL

Mass of small filter paper: 0.16 g

Mass of watch glass: 47.19 g

Mass of watch glass + filter + product: 47.44 g

Final mass of product: 0.09 g

Physical state and color of product: flaky white solid

experimental melting point: 58-64°

TLC diagram:

Lit MP:

R_f

dibromo: $\dfrac{0.7}{4.7} = 1.5$ 0.7 59-61°

our sample: $\dfrac{1.1}{4.7} = 0.23$

diphenylacetate: $\dfrac{1}{4.7} = .21$

Theoretical yield calculation based on both reactants

$$\dfrac{0.47}{\text{dibromo}} \left| \dfrac{1\,\text{mol rct}}{340.06\,g} \right| \dfrac{4\,\text{mol prod}}{1\,\text{mol rct}} \left| \dfrac{178.23\,g}{1\,\text{mol prod}} \right. = 0.246\ g\ \text{diphenyl acetylene}$$

$$0.42\ g\ KOH \left| \dfrac{1\,\text{mol rct}}{56\,g} \right| \dfrac{1\,\text{mol prod}}{1\,\text{mol rct}} \left| \dfrac{178.23\,g}{1\,\text{mol prod}} \right. = 1.337\ g\ \text{diphenyl acetylene}$$

Percent yield calculation:

$$\dfrac{.09\ g}{0.246\ g} \times 100\% = 36.59\%$$

Experiment title and number

Date

Name

Course

Section

Lab partner

Experiment title and number

Date

Name

Name   Course   Section   Lab partner

Caution: Place fold-in flap under yellow sheet before writing, to protect the pages that follow.

Experiment title and number

Date

Name

Course

Section

Lab partner

Experiment title and number

Date

Name

Course

Section

Lab partner

Caution: Place fold-in flap under yellow sheet before writing, to protect the pages that follow.

Experiment title and number		Date

Name	Course	Section	Lab partner

Experiment title and number		Date	

Name	Course	Section	Lab partner

Experiment title and number

Date

Name

Course

Section

Lab partner

Experiment title and number

Date

Caution: Place fold-in flap under yellow sheet before writing, to protect the pages that follow.

Name

Course

Section

Lab partner

Name

Course

Section

Lab partner

Experiment title and number

Date

16

Name

Course

Section

Lab partner

Caution: Place fold-in flap under yellow sheet before writing, to protect the pages that follow.

Experiment title and number

Date

17

Name

Course

Section

Lab partner

Experiment title and number		Date	
Name	Course	Section	Lab partner

Experiment title and number

Date

Name

Course

Section

Lab partner

Experiment title and number

Date

Name

Course

Section

Lab partner

Caution: Place fold-in flap under yellow sheet before writing, to protect the pages that follow.

Name

Course

Section

Lab partner

Name

Course

Section

Lab partner

Experiment title and number		Date
Name	Course	Section	Lab partner

Caution: Place fold-in flap under yellow sheet before writing, to protect the pages that follow.

Experiment title and number		Date

Name	Course	Section	Lab partner

Caution: Place fold-in flap under yellow sheet before writing, to protect the pages that follow.

Experiment title and number

Date

22

Name

Course

Section

Lab partner

Experiment title and number

Date

Name

Course

Section

Lab partner

Caution: Place fold-in flap under yellow sheet before writing, to protect the pages that follow.

Experiment title and number				Date		
Name			Course	Section	Lab partner	

Experiment title and number				Date		

Experiment title and number

Date

Name

Course

Section

Lab partner

Experiment title and number

Date

Caution: Place fold-in flap under yellow sheet before writing, to protect the pages that follow.

Experiment title and number

Date

Name

Course

Section

Lab partner

Experiment title and number

Date

Name

Caution: Place fold-in flap under yellow sheet before writing, to protect the pages that follow.

Experiment title and number		Date	

Name		Course	Section	Lab partner

Experiment title and number

Date

25

Name

Course

Section

Lab partner

Caution: Place fold-in flap under yellow sheet before writing, to protect the pages that follow.

Experiment title and number		Date	

Name	Course	Section	Lab partner

Experiment title and number

Date

Name

Course

Section

Lab partner

Caution: Place fold-in flap under yellow sheet before writing, to protect the pages that follow.

Experiment title and number

Date

27

Name

Course

Section

Lab partner

Experiment title and number

Date

Name

Course

Section

Lab partner

Caution: Place fold-in flap under yellow sheet before writing, to protect the pages that follow.

Experiment title and number		Date	

Name	Course	Section	Lab partner

Experiment title and number		Date

Name	Course	Section	Lab partner

Caution: Place fold-in flap under yellow sheet before writing, to protect the pages that follow.

Experiment title and number				Date	
Name		Course	Section	Lab partner	

Name	Course	Section	Lab partner

Caution: Place fold-in flap under yellow sheet before writing, to protect the pages that follow.

Name | Course | Section | Lab partner

Name | Course | Section | Lab partner

Caution: Place fold-in flap under yellow sheet before writing, to protect the pages that follow.

Experiment title and number		Date	
Name		Course	Section	Lab partner

Experiment title and number		Date	
Name	Course	Section	Lab partner

Name

Course

Section

Lab partner

Caution: Place fold-in flap under yellow sheet before writing, to protect the pages that follow.

Experiment title and number

Date

Name

Course

Section

Lab partner

Experiment title and number		Date		
Name		Course	Section	Lab partner

Name

Course

Section

Lab partner

Caution: Place fold-in flap under yellow sheet before writing, to protect the pages that follow.

Name | Course | Section | Lab partner

Caution: Place fold-in flap under yellow sheet before writing, to protect the pages that follow.

Name | Course | Section | Lab partner

Caution: Place fold-in flap under yellow sheet before writing, to protect the pages that follow.

Experiment title and number

Date

Name

Course

Section

Lab partner

Caution: Place fold-in flap under yellow sheet before writing, to protect the pages that follow.

Experiment title and number

Date

Name

Course

Section

Lab partner

Caution: Place fold-in flap under yellow sheet before writing, to protect the pages that follow.

Experiment title and number		Date	
Name	Course	Section	Lab partner

Experiment title and number

Date

Name

Course

Section

Lab partner

Experiment title and number		Date	
Name	Course	Section	Lab partner

Name

Course

Section

Lab partner

Name

Course

Section

Lab partner

Caution: Place fold-in flap under yellow sheet before writing, to protect the pages that follow.

Experiment title and number

Date

42

Name

Course

Section

Lab partner

Caution: Place fold-in flap under yellow sheet before writing, to protect the pages that follow.

Experiment title and number		Date
Name	Course	Section	Lab partner

Caution: Place fold-in flap under yellow sheet before writing, to protect the pages that follow.

Experiment title and number		Date	43
Name	Course	Section	Lab partner

Experiment title and number		Date	
Name	Course	Section	Lab partner

Experiment title and number		Date	

Name	Course	Section	Lab partner

Caution: Place fold-in flap under yellow sheet before writing, to protect the pages that follow.

Experiment title and number

Date

45

Name

Course

Section

Lab partner

Caution: Place fold-in flap under yellow sheet before writing, to protect the pages that follow.

Experiment title and number

Date

Name

Course

Section

Lab partner

Caution: Place fold-in flap under yellow sheet before writing, to protect the pages that follow.

Experiment title and number

Date

46

Name

Course

Section

Lab partner

Experiment title and number

Date

46

Name

Course

Section

Lab partner

Caution: Place fold-in flap under yellow sheet before writing, to protect the pages that follow.

Caution: Place fold-in flap under yellow sheet before writing, to protect the pages that follow.

Caution: Place fold-in flap under yellow sheet before writing, to protect the pages that follow.

Experiment title and number

Date

48

Name

Course

Section

Lab partner

Caution: Place fold-in flap under yellow sheet before writing, to protect the pages that follow.

Experiment title and number   Date

Name   Course   Section   Lab partner

Experiment title and number		Date	
Name	Course	Section	Lab partner

Experiment title and number

Date

Name

Course

Section

Lab partner

Experiment title and number

Date

Caution: Place fold-in flap under yellow sheet before writing, to protect the pages that follow.

Experiment title and number

Date

Name

Course

Section

Lab partner

Name | Course | Section | Lab partner

Experiment title and number

Date

50

Experiment title and number		Date	51
Name	Course	Section	Lab partner

Caution: Place fold-in flap under yellow sheet before writing, to protect the pages that follow.

Name

Course

Section

Lab partner

Experiment title and number

Date

52

Name

Course

Section

Lab partner

Name   Course   Section   Lab partner

Caution: Place fold-in flap under yellow sheet before writing, to protect the pages that follow.

Caution: Place fold-in flap under yellow sheet before writing, to protect the pages that follow.

| Name | | Course | Section | Lab partner |

Name | Course | Section | Lab partner

Name

Course

Section

Lab partner

Experiment title and number

Date

57

Name

Course

Section

Lab partner

Experiment title and number

Date

Name

Course

Section

Lab partner

Name | Course | Section | Lab partner

Caution: Place fold-in flap under yellow sheet before writing, to protect the pages that follow.

Experiment title and number		Date

Name		Course	Section	Lab partner

Caution: Place fold-in flap under yellow sheet before writing, to protect the pages that follow.

Experiment title and number

Date

Name

Bottles Section Lab partner

Experiment title and number

Date

Name

Course

Section

Lab partner

Experiment title and number

Date

Name

Caution: Place fold-in flap under yellow sheet before writing, to protect the pages that follow.

Name

Course

Section

Lab partner

Caution: Place fold-in flap under yellow sheet before writing, to protect the pages that follow.

Name | Course | Section | Lab partner

Name | Course | Section | Lab partner

Caution: Place fold-in flap under yellow sheet before writing, to protect the pages that follow.

Experiment title and number

Date

62

Name

Course

Section

Lab partner

Name

Course

Section

Lab partner

Caution: Place fold-in flap under yellow sheet before writing, to protect the pages that follow.

Name

Course

Section

Lab partner

Experiment title and number				Date	
Name			Course	Section	Lab partner

Name | Course | Section | Lab partner

Caution: Place fold-in flap under yellow sheet before writing, to protect the pages that follow.

Experiment title and number

Date

65

Name

Course

Section

Lab partner

Experiment title and number		Date

Name		Course	Section	Lab partner

Caution: Place fold-in flap under yellow sheet before writing, to protect the pages that follow.

Name

Course

Section

Lab partner

Caution: Place fold-in flap under yellow sheet before writing, to protect the pages that follow.

Name | Course | Section | Lab partner

Caution: Place fold-in flap under yellow sheet before writing, to protect the pages that follow.

Name	Course	Section	Lab partner

Caution: Place fold-in flap under yellow sheet before writing, to protect the pages that follow.

Experiment title and number				Date	
Name			Course	Section	Lab partner

Experiment title and number		Date

Name	Course	Section	Lab partner

Caution: Place fold-in flap under yellow sheet before writing, to protect the pages that follow.

Caution: Place fold-in flap under yellow sheet before writing, to protect the pages that follow.

Experiment title and number		Date
Name	Course	Section	Lab partner

Caution: Place fold-in flap under yellow sheet before writing, to protect the pages that follow.

Experiment title and number		Date	
Name	Course	Section	Lab partner

Caution: Place fold-in flap under yellow sheet before writing, to protect the pages that follow.

Name   Course   Section   Lab partner

Caution: Place fold-in flap under yellow sheet before writing, to protect the pages that follow.

Experiment title and number

Date

72

Name

Course

Section

Lab partner

Experiment title and number

Date

Name

Course

Section

Lab partner

Caution: Place fold-in flap under yellow sheet before writing, to protect the pages that follow.

Experiment title and number

Date

73

Name

Course

Section

Lab partner

Experiment title and number		Date	
Name	Course	Section	Lab partner

Caution: Place fold-in flap under yellow sheet before writing, to protect the pages that follow.

Name

Course

Section

Lab partner

Experiment title and number

Date

Name

Course

Section

Lab partner

Caution: Place fold-in flap under yellow sheet before writing, to protect the pages that follow.

Experiment title and number		Date	
Name	Course	Section	Lab partner

Name		Course	Section	Lab partner

Experiment title and number		Date	

Caution: Place fold-in flap under yellow sheet before writing, to protect the pages that follow.

Experiment title and number

Date

Name

Course

Section

Lab partner

Caution: Place fold-in flap under yellow sheet before writing, to protect the pages that follow.

Experiment title and number			Date		
Name		Course	Section	Lab partner	

Caution: Place fold-in flap under yellow sheet before writing, to protect the pages that follow.

Name | Course | Section | Lab partner

Experiment title and number		Date	
Name	Course	Section	Lab partner

Experiment title and number				Date	
Name			Course	Section	Lab partner

Experiment title and number				Date	
Name			Course	Section	Lab partner

Experiment title and number

Date

80

Name

Course

Section

Lab partner

Name | Course | Section | Lab partner

Experiment title and number

Date

81

Name

Course

Section

Lab partner

Experiment title and number

Date

Name

Course

Section

Lab partner

Caution: Place fold-in flap under yellow sheet before writing, to protect the pages that follow.

Experiment title and number		Date		
Name	Course	Section	Lab partner	

Name

Course

Section

Lab partner

Caution: Place fold-in flap under yellow sheet before writing, to protect the pages that follow.

Experiment title and number

Date

83

Name

Course

Section

Lab partner

Caution: Place fold-in flap under yellow sheet before writing, to protect the pages that follow.

Experiment title and number

Date

Name

Course

Section

Lab partner

Caution: Place fold-in flap under yellow sheet before writing, to protect the pages that follow.

Experiment title and number		Date	

Name		Course	Section	Lab partner

Experiment title and number		Date	

Caution: Place fold-in flap under yellow sheet before writing, to protect the pages that follow.

Experiment title and number		Date	
Name	Course	Section	Lab partner

Caution: Place fold-in flap under yellow sheet before writing, to protect the pages that follow.

Experiment title and number

Date

Name

Course

Section

Lab partner

Experiment title and number				Date	
Name			Course	Section	Lab partner

Experiment title and number		Date

Name		Course	Section	Lab partner

Experiment title and number		Date	
Name	Course	Section	Lab partner

Caution: Place fold-in flap under yellow sheet before writing, to protect the pages that follow.

Experiment title and number

Date

88

Name

Course

Section

Lab partner

Experiment title and number		Date	

Name		Course	Section	Lab partner

Experiment title and number		Date

Name		Course	Section	Lab partner

Caution: Place fold-in flap under yellow sheet before writing, to protect the pages that follow.

Experiment title and number

Date

Name

Course

Section

Lab partner

Experiment title and number		Date

Name	Course	Section	Lab partner

Experiment title and number			Date	
Name		Course	Section	Lab partner

Experiment title and number

Date

92

Name

Course

Section

Lab partner

Experiment title and number

Date

Name

Course

Section

Lab partner

Caution: Place fold-in flap under yellow sheet before writing, to protect the pages that follow.

Experiment title and number | Date | 93

Name | Course | Section | Lab partner

| Name | Course | Section | Lab partner |

Experiment title and number

Date

94

Name

Course

Section

Lab partner

Caution: Place fold-in flap under yellow sheet before writing, to protect the pages that follow.

Name | Course | Section | Lab partner

Experiment title and number

Date

Name

Course

Section

Lab partner

Experiment title and number		Date	

Name		Course	Section	Lab partner

Caution: Place fold-in flap under yellow sheet before writing, to protect the pages that follow.

Experiment title and number

Date

Name

Course

Section

Lab partner

Name

Course

Section

Lab partner

Caution: Place fold-in flap under yellow sheet before writing, to protect the pages that follow.

Experiment title and number

Date

Name

Course

Section

Lab partner

Caution: Place fold-in flap under yellow sheet before writing, to protect the pages that follow.

Experiment title and number

Date

99

Name

Course

Section

Lab partner

Experiment title and number		Date	
Name	Course	Section	Lab partner

Caution: Place fold-in flap under yellow sheet before writing, to protect the pages that follow.

Experiment title and number		Date

Name	Course	Section	Lab partner